FOCUS ON

ELEMENTARY

Grades K-4

GEOLOGY

Rebecca W. Keller, PhD

Illustrations: Janet Moneymaker
Marjie Bassler

Focus On Elementary Geology Student Textbook (hardcover)
ISBN: 978-1-936114-88-7

Published by Gravitas Publications, Inc.
www.gravitaspublications.com

CONTENTS

iv Contents

Chapter 1: What Is Geology?

1.1 Introduction

Do you ever pick up rocks and wonder how they were made and what they are made of? Do you sometimes look at mountains and wonder how they were formed? Have you wondered what's at the bottom of the ocean? Have you noticed how weather affects the landscape? Do you wonder why certain birds and wild animals live near you and others don't?

These are all questions that are explored by scientists who study geology. Geology is the study of the Earth. By studying the Earth, scientists attempt to understand what the Earth is made of, how Earth came into being, how Earth has changed in the past, how it is changing now, and our role as we live on Earth.

1.2 History of Geology

Depending on where you live, when you go outside and walk around, you'll see many different features of Earth. You might see mountains or rivers. You might

see fields of dirt or fields of grass. You might see lakes or oceans, mesas or glaciers, forests or prairies.

Ancient people also saw many of the same features you see. Although lakes come

and go and rivers might change course, many of the features you see today are the same features ancient people would have seen.

One of the very first people to study Earth's features was the Greek philosopher Theophrastus who lived from about 371-287 BCE.

Theophrastus was a student of Aristotle who was another famous Greek philosopher. Like Aristotle, Theophrastus was interested in science. He studied rocks and explored what happened when rocks were heated.

Many of the first geologists also asked questions about how the Earth came into being and how many years the Earth has existed. All of these questions shaped the modern science we now call geology.

1.3 Modern Geology

Modern geologists continue to study rocks and what they are made of, and they ask questions about how mountains and glaciers form. Modern geologists have an advantage over ancient geologists because modern geologists can use chemistry and physics to better understand how things work.

Some modern geologists focus on the chemistry of Earth. These geologists are called geochemists. Geochemists study how atoms and molecules make rocks, soils, minerals, and fuels.

Other modern geologists focus on the structure of Earth. These geologists are called structural geologists. Structural geologists study how Earth is put together and how it changes. They are interested in how rocks change and what makes mountains and valleys.

Other modern geologists study how humans affect the water, air, and land quality of Earth. These geologists study Earth's environment and are called environmental geologists.

1.4 Everyday Geology

Even though you may not be a geologist yet, you can learn about the Earth by simply observing what happens around you.

What happens when it rains? Do the roads flood? Do you get mud slides, or does a river find a new path? What happens in the hot sun? Do you observe mud forming cracks or rocks crumbling? Have you ever been in an earthquake? Did you feel the ground move?

Paying attention to where you live, what happens during storms, and how the land around you changes over time are activities you can do every day.

1.5 Summary

● Geology is the study of Earth.

● The first geologists looked at rocks and minerals and asked questions about how Earth came into being.

● Geochemists are modern geologists who study how atoms and molecules form Earth.

● Structural geologists look at how Earth is put together.

● Environmental geologists look at changes in the quality of the water, the air, and the land on Earth.

Chapter 2: Hammers and Lenses

2.1 Introduction

In Chapter 1 we saw that geology is the study of Earth. We also saw that modern geology includes different areas of study, such as geochemistry and structural geology.

Scientists in each of these areas of geology may use different tools. A geochemist might use a test tube to study what rocks are made of, and a structural geologist might use a seismograph, which is an instrument that records motions below the surface of the Earth.

However, all geologists use two common tools: rock hammers and hand lenses.

2.2 Types of Hammers

Three types of rock hammers used by geologists are: the pointed tip rock hammer, the chisel edge rock hammer, and the crack hammer.

A pointed tip rock hammer has one end that comes to a point. The other end has a head that is flat and square. Pointed tip rock hammers are also called rock picks, and they are generally used by geologists when they are working with hard rocks. Geologists often use the pointy tip to dig fossil samples out of rocks and the square end to crack open a rock to see what's inside.

A chisel edge rock hammer has one end that is flat and broad like a chisel rather than being pointy. The other end is a square head. The chisel end is used to split layers of soft rock. The square head is used to crack open rocks.

A crack hammer has two blunt ends. It is usually heavier than either a pointed tip rock hammer or a chisel edge rock hammer. Its heaviness can make it easier for geologists to break rocks open.

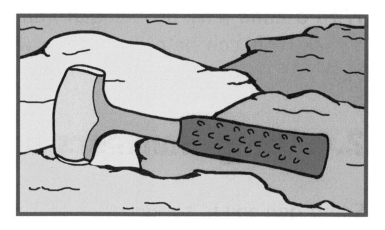

2.3 Hand Lenses

Once a geologist breaks a rock open with a rock hammer, is there a way to get a better look at what's inside? To better see the features of the inside of a rock, geologists often use a hand lens. A hand lens is a small magnifier that folds into a holder, making it compact and easy to carry. A hand lens can be carried in a pocket or on a string around the neck.

A hand lens magnifies a rock, making it easier to see. With a hand lens, a geologist can look at the details of a rock. This can help determine what the rock is made of.

2.4 Using Hammers and Lenses

Some geologists spend a lot of time outdoors. They might go hiking, or they might climb up mountains.

Even though they need to bring tools to help them study the rocks they find, they need their tools to be easy to carry. Rock hammers and hand lenses are ideal tools for geologists because they are usually small and lightweight.

Even though rock hammers and hand lenses are simple tools, geologists need to learn how use them effectively. They can't tap a rock too lightly or it might not break open. They also can't hit a rock too hard or it might crumble, destroying delicate structures. Learning how to tap a rock just right is a skill geologists learn over time.

Learning how to focus a hand lens also takes some time. If the lens is held too close to the rock, it won't magnify. If the lens is too far away, it won't focus and everything will look fuzzy.

2.5 Summary

● Geologists have different tools that they use.

● Many geologists use rock hammers and hand lenses.

● Three types of rock hammers used by geologists are: the pointed tip rock hammer, the chisel edge rock hammer, and the crack hammer.

● Hand lenses help geologists see what is inside rocks.

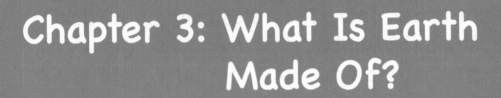

Chapter 3: What Is Earth Made Of?

3.1 Introduction

If you walk outside, do you notice where you live? If you live in the country, do you notice the trees and grass? If you live in a city, do you notice the buildings and streets?

If you look down, do you notice the ground? If you live in a city, do you notice how much of the ground is covered up with streets or pavement? If you live in the country, do you notice if the ground has grass, rocks, or dirt?

When you notice the ground with the rocks, dirt, grass, and trees, you are noticing the Earth. The Earth is where you live.

But what is the Earth made of? What is dirt? What are rocks? Why do grass and trees grow in dirt? How deep does the dirt go? How many kinds of rocks are there? What is below the rocks and dirt? More rocks? Trees? Chocolate syrup?

In this chapter we will learn about what the Earth is made of.

3.2 Rocks and Minerals

The crust is the outer part of Earth and is where we live. The crust is mostly made of rock. If you go outside and start digging in the ground with a shovel, you will eventually hit some type of rock.

What is the difference between a rock and a tree? Trees are living things. Rocks are not living things. Rocks do not move like living things. Rocks do not grow like living things, and rocks do not multiply like living things.

Rocks are made mostly of silicon, and living things are made mostly of carbon. Silicon and carbon are atoms (also called elements). Because rocks are made mostly from silicon and living things are made mostly from carbon, rocks are very different from living things.

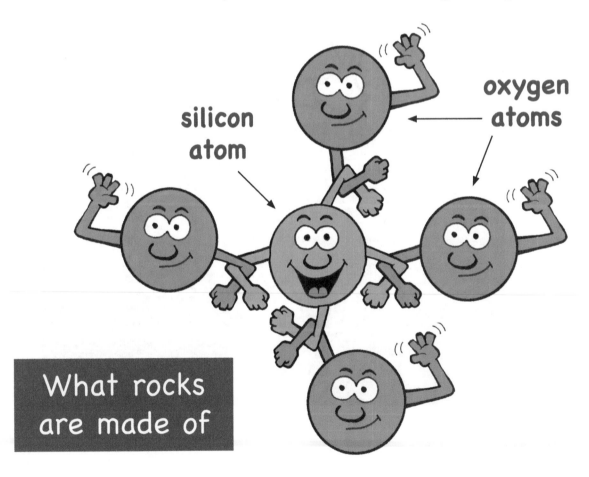

silicon atom

oxygen atoms

What rocks are made of

All rocks come from magma, which is molten (melted) rock deep inside the Earth. Magma is made mostly of the elements silicon and oxygen. Rocks form when the magma cools and mixes with other elements, like magnesium, iron, or aluminum.

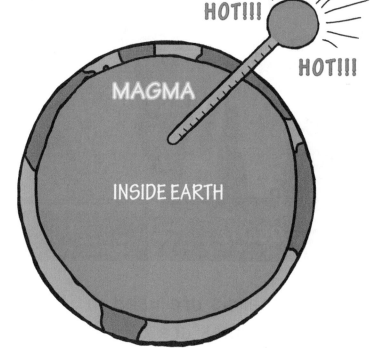

When magma cools very slowly, the atoms in the magma have a chance to line up in an orderly fashion. When this happens, the material that is formed is called a mineral.

There are many different kinds of minerals. One common mineral found in rocks is quartz. Quartz can be clear, pink, purple, or other colors.

Mica is also a mineral. Mica is soft and looks like thin, layered paper. You can peel mica sheets away from each other.

Calcite is another mineral found in rocks. Calcite is made of calcium and oxygen and forms beautiful crystals that come in different colors.

Minerals

Quartz   Mica   Calcite

Some minerals are used in jewelry. Rubies are a brilliant red-colored mineral made of aluminum and oxygen. Emeralds are a different type of mineral made of aluminum, beryllium, silicon, and oxygen. Emeralds have a deep green color. Both rubies and emeralds are hard to find, and that is why jewelry made with them is often expensive.

Ruby Crystal

There are many different kinds of rocks. There are rocks that are formed deep inside the Earth. There are other rocks that form on the surface of the Earth. And there are still other rocks that have been changed from one type of rock into another type of rock. The different types of rocks are called igneous, sedimentary, and metamorphic. Minerals are the building blocks of all these types of rocks.

Types of Rock

Igneous
(granite)

Sedimentary
(limestone)

Metamorphic
(garnet schist)

Igneous rocks are the most plentiful type of rock. Igneous rocks are formed when molten magma inside the Earth cools and hardens. Granite is one kind of igneous rock, and it has a lot of quartz in it.

Sedimentary rocks are formed from bits of rocks and other materials that are left behind by wind or water. As more and more of these materials pile on top of each other in one area, the layers underneath are pressed together. This pressure turns the materials to rock.

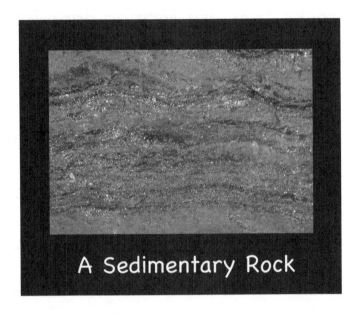

A Sedimentary Rock

A Metamorphic Rock

Metamorphic rocks are those that have changed from one type of rock into another. Very high heat and pressure cause these changes. So igneous rocks and sedimentary rocks can be changed into metamorphic rocks. Even metamorphic rocks can be changed into other metamorphic rocks!

3.3 Dirt

When you go outside and dig a hole in the ground, you not only find rocks, but you also find dirt. Dirt is made of rocks, plants, minerals, and even small animals!

There are many different types of dirt. Some dirt is sandy and light in color. Other dirt is moist and dark in color. Have you ever wondered why some parts of the world are used for growing food and other parts of the world are not? That's because some dirt is good for growing plants and other dirt is not.

Dirt is also called soil, and it comes mainly from rocks. Since different parts of the Earth have different kinds of rocks, there will be different types of soil in different places.

3.4 Summary

● The Earth is made mostly of rock.

● All rocks come from magma.

● Minerals form when magma cools slowly, allowing the atoms to line up in an orderly fashion.

● Dirt is called soil. It is made mostly from rocks and contains materials that come from plants, minerals, and animals.

Chapter 4: Our Earth

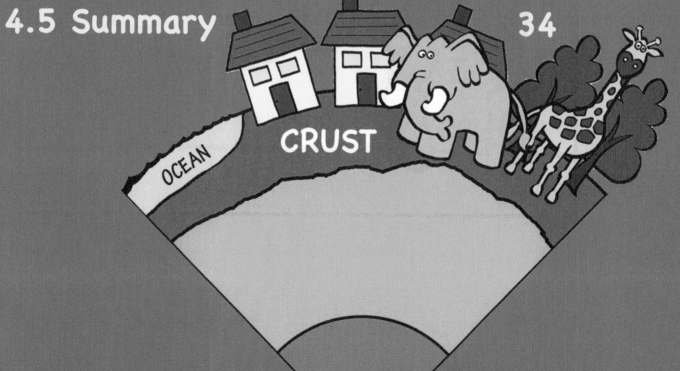

OCEAN   CRUST

4.1 Shape of Earth

What is the shape of Earth? Is it round like a circle? Is it spherical like a ball? Is it flat like a pancake? Is it square like a block? Does it look like a pumpkin or an eggplant?

When you take a walk around your neighborhood, Earth seems flat. As you go to the park or walk to the store, you don't slide off the ground and you don't lean to the left or right. When you walk on a hill, you go up and then down again. You can walk for many miles and the Earth will seem flat.

However, if you change where you are and look at the Earth from an airplane or from a boat at sea, you'll discover that the Earth is not flat, but curved. If you looked at Earth from a spaceship, you would discover that Earth is shaped like a ball.

But Earth is not shaped like a perfectly round ball. The middle of the Earth is pushed out a little, and the top and bottom are slightly flattened. The Earth is shaped like a slightly smashed ball!

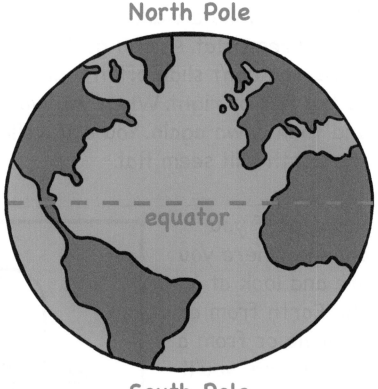

The top part of Earth is called the North Pole, and the bottom part of Earth is called the South Pole. The middle of Earth is called the equator.

4.2 Size of Earth

It's hard to imagine that we walk, run, and live on a planet that's shaped like a ball. If you live at the middle of Earth, near the equator, do you feel like you are standing sideways? If you go to the South Pole do you feel upside down? If you go to the North Pole do you feel right-side up?

In fact, if you go anywhere on the Earth, you don't feel upside down or sideways. You always feel right-side up.

Why does Earth seem flat if it is actually a curved ball? And why do we always feel upright even when we are at the South Pole or on the equator?

Many years ago some people were afraid to travel too far out to sea. They thought the Earth was flat and that they would fall off the edge if they went too far! It made sense to them because the Earth felt flat.

What they didn't know is that the Earth is very large. The Earth is so large that we can't feel the Earth's curve when we walk or run. The Earth is so large that we can't tell if we are on the top, on the bottom, or on the side.

The Earth is about 25,000 miles around. It would take you more than a year to walk around the Earth! That's how big Earth is.

Because Earth is so large, it feels flat when we walk, run, work, or play. But the Earth is really a huge ball.

4.3 Parts of Earth

Earth has different parts. You know that when you walk outside, you step on rocks and dirt. Rocks, minerals, and dirt make up the outer part of Earth. Rocks, minerals, and dirt also make up the ocean floor.

But what is beneath the dirt and the ocean? What is inside the Earth? Is it rocks, minerals, and dirt like the outer part? Or is it chocolate syrup, liquid gold, or melted cheese?

Scientists have no way to actually see what's below the outer surface of Earth, but by studying volcanoes and earthquakes, scientists can come up with some ideas about what lies below the surface.

4.4 Earth's Layers

Earth is a rock planet, which means that it is made mostly of rocks.

If scientists were able to cut the Earth in half, they think they would see at least three different layers. This means that the outer surface of Earth, where you walk, is different from the part just below it.

The crust is the outermost layer of Earth. The crust is made up of rocks, minerals, and soil. The crust is very hard. This hard outer layer supports you when you walk and supports buildings. The crust also holds the oceans. The crust is relatively thin compared to the layers below it, and it makes up only a small part of Earth.

Below the crust is the mantle. Scientists think the mantle is much thicker than the crust and is made of layers. Scientists believe the outer mantle is hard and rocky and the inner mantle is softer and hotter and

might be more like gooey peanut butter. The inner mantle is made of melted rock called magma. Scientists think that the magma in the mantle does not always stay in the same place but moves around!

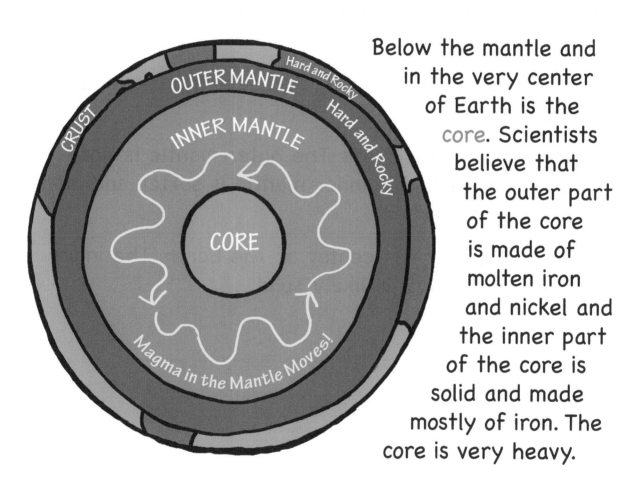

Below the mantle and in the very center of Earth is the core. Scientists believe that the outer part of the core is made of molten iron and nickel and the inner part of the core is solid and made mostly of iron. The core is very heavy.

Earth's crust, mantle, and core together are called the geosphere.

4.5 Summary

● The Earth is made of layers.

● The main layers of the Earth are the crust, the mantle, and the core.

● The crust is the hard, rocky, outermost layer of Earth.

● The mantle has layers. The outer mantle is hard and rocky, and the inner mantle is softer and more like peanut butter.

● The core is at the center of the Earth. The inner part of the core is likely solid.

Chapter 5: Earth Is Active

5.1 Introduction

The Earth is an active planet. It is always changing. For example, we can see the surface of the Earth change with the seasons. In the spring and summer some parts of Earth can be covered with flowers and

green grass. In the winter, these same parts might be covered with snow.

You might think that the Earth's crust doesn't change because it is hard and rocky, but it changes every day. There are volcanoes, earthquakes, storms, and flowing rivers that continuously change the Earth's crust.

Geologists study how Earth has changed by looking at rocks, dirt layers, mountains, and other features. Geologists also look at volcanoes, earthquakes, weather, and other things that cause changes on Earth.

5.2 Volcanoes Erupt

Volcanoes can be very exciting. Sometimes when a

volcano erupts, a whole mountaintop will come off in a huge explosion!

Volcanoes happen when magma in the mantle pushes up through a weak spot in the crust and comes to the surface of the Earth.

Magma is formed when the rocks and minerals in the mantle melt because the weight of the crust is pushing down on them, creating heat and pressure.

Pressure happens when you squeeze something and it doesn't have anywhere to go. For example, if you keep the lid on your toothpaste and squeeze the bottom of the tube, you will create pressure in the tube. If the top suddenly pops off, the toothpaste will explode! If there is a crack in the wall of the toothpaste tube, the toothpaste will squirt out the side.

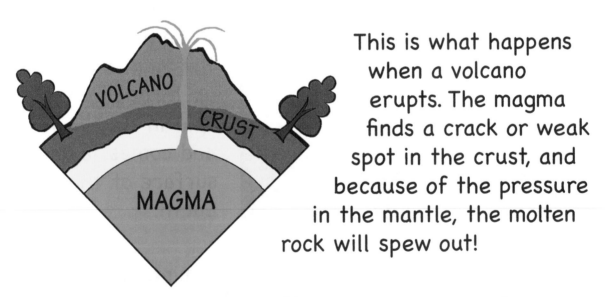

This is what happens when a volcano erupts. The magma finds a crack or weak spot in the crust, and because of the pressure in the mantle, the molten rock will spew out!

When the magma comes out onto the surface of the Earth, it is called lava. Lava that has flowed on Earth's surface is responsible for many interesting features. Sometimes lava flows form whole islands. The Hawaiian Islands, for example, are a group of islands made from lava flows.

Volcanoes can form mountains of different shapes and sizes. Some volcanoes form gently sloping mountains over long periods of time. These volcanoes are called shield volcanoes because their long slopping sides make

them resemble the shape of a shield. Shield volcanoes form when very thin layers of lava flow out in all directions. A mountain is formed as the layers build up on top of each other. Shield volcanoes can be several miles long with sides that slope very gradually. The Hawaiian Islands are a series of shield volcanoes.

Volcanoes can also form mountains with steeper sides. Cone volcanoes are cone-shaped because the magma spurts out more quickly and is thicker than the magma that forms shield volcanoes. Also, more rocks and dirt are scattered. The rocks and dirt pile up along the sides of the volcano, making the sides steeper and steeper. Interestingly, cone volcanoes are often found on the edges of shield volcanoes.

Volcanoes can also form dome-shaped mountains. Lava dome mountains are often round in shape and look like a cereal bowl turned upside down! The magma that comes from dome volcanoes is very thick and doesn't flow very far away from the center.

5.3 Earthquakes Shake

Earthquakes can also be very exciting and even scary if you happen to be near one when it happens. Earthquakes occur because sections of the Earth's crust suddenly move. Recall that the hard rocky crust lies on top of the outer mantle, and the hard outer mantle lies on top of the inner mantle, which is soft like peanut butter.

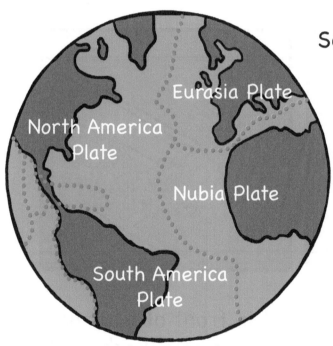

Scientists think that the hard outer mantle is cracked into huge pieces called plates that fit together like a big puzzle. These pieces or plates float on the part of the mantle that is like peanut butter.

Earthquakes happen suddenly and are usually over within a few minutes. If the earthquake is small, you might feel the floor move, and you might hear a sound like a nearby train. If the earthquake is large, the whole ground moves and buildings and trees may fall. Sometimes after a big earthquake, several smaller earthquakes occur.

When plates move against each other, one section might move upward and the other might move downward. Or the sections might slide past one another. When these things happen, there is movement of the land on either side of the line that is between the two sections.

Let's imagine you have a street in front of your house that separates your yard from your neighbor's yard. Imagine that your house sits on one section of the Earth and your neighbor's house sits on a different section of the Earth. If these two sections move past one another, it's possible that your neighbor from across the street is now your next-door neighbor!

5.4 Summary

● The Earth's surface is constantly changing.

● Volcanoes and earthquakes change Earth's crust.

● Volcanoes happen when magma in the mantle pushes up through a weak spot in the Earth's surface.

● Earthquakes happen when sections of the Earth's surface move.

Chapter 6: The Air We Breathe

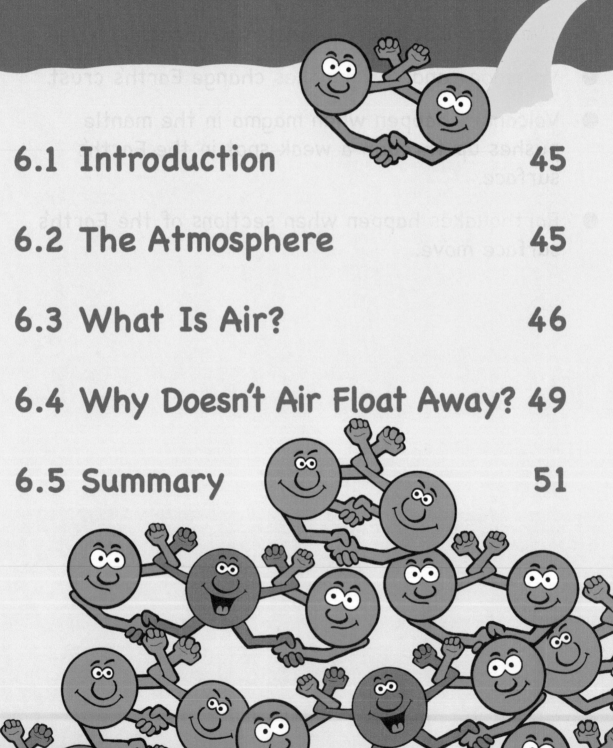

6.1 Introduction

Notice what happens when you breathe. Take a deep breath in. What is going into your lungs? Exhale. What comes out?

Air is the name for what we breathe in and what we breathe out. We live on a planet that has the kind of air needed for life. Without the air on Earth, there wouldn't be animals on farms, ponds full of fish, or forests full of trees.

6.2 The Atmosphere

The air we breathe exists in the part of the Earth that scientists call the atmosphere. The atmosphere sits just above the Earth's crust and extends for

several miles above the surface. Earth is the only planet in our solar system that has an atmosphere suitable for life as we know it.

Most of the time we don't think too much about the air that surrounds us. We breathe it in and we exhale it out as we go about our day. Some people, though, live in areas where the air is not clean. In some areas of the world, the air contains so much pollution, or small particles, that it is difficult to breathe. Knowing about the air, what it is, and how to keep it clean is important to all living things on Earth.

6.3 What Is Air?

You might think most of the air we breathe is oxygen. But it isn't. The air we breathe is a mixture of different gas molecules and water vapor. Air has nitrogen gas, oxygen gas, carbon dioxide gas, and a little bit of argon gas. Nitrogen gas in the air exists as a molecule with two nitrogen atoms hooked together. When you breathe, nitrogen gas goes inside your lungs and

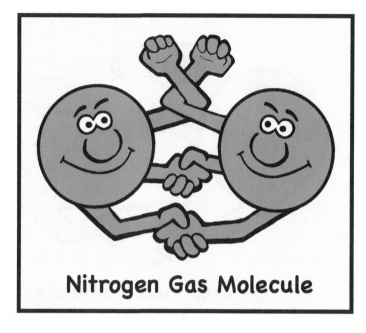

Nitrogen Gas Molecule

comes out again when you exhale. Your body doesn't use nitrogen gas. It only uses oxygen gas.

You have probably heard that you need oxygen to stay alive. In fact, if you hold your breath too long, you will pass out and your body will automatically start breathing again to keep you alive. Oxygen is more important than food or water, and you can't stay alive very long without it.

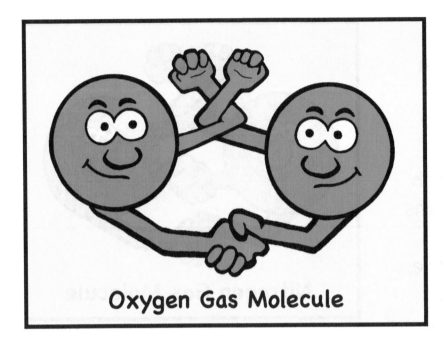

Oxygen Gas Molecule

Oxygen gas is made of two oxygen atoms hooked together. When you breathe in, your lungs expand and oxygen gas goes inside.

Lungs have special cells that absorb oxygen from the air and deliver it to your blood. Your blood has special molecules that then carry the oxygen to the rest of your body. Your body uses the oxygen to process food and give you energy.

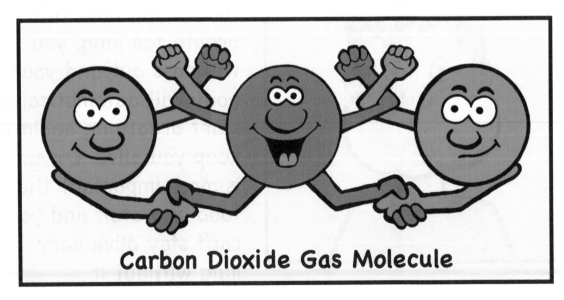

Carbon Dioxide Gas Molecule

After our bodies use the oxygen we have inhaled, they make carbon dioxide gas. Carbon dioxide gas is made of one carbon atom and two oxygen atoms hooked together. When you exhale, you breathe out the carbon dioxide gas.

This works out great because plants use carbon dioxide to make the food that allows them to stay alive and grow. Plants take in carbon dioxide and then put oxygen back into the air. We help plants and plants help us!

6.4 Why Doesn't Air Float Away?

You might have wondered if there is oxygen, nitrogen, or carbon dioxide in space. In fact, when astronauts travel outside our atmosphere and into space, they have to wear a special suit so that they can have oxygen to breathe.

But what keeps the air we breathe close to the Earth, and why doesn't it just all float away into space? In fact, what keeps you on the Earth, and why don't you float away?

You don't float away for the same reason the air in our atmosphere doesn't float away. Gravity keeps you on the ground and keeps the air we breathe close to the Earth's surface. Gravity is a force that pulls everything near the Earth toward its center. Go ahead and try to jump off the Earth. As soon as you jump up, you will feel Earth's gravity pulling you back down again.

6.5 Summary

● The air we breathe is in the Earth's atmosphere.

● The air we breathe is made up of nitrogen gas, oxygen gas, carbon dioxide gas, argon gas, and water vapor.

● Gravity keeps the air from floating away into space.

Chapter 7: Our Water

7.1 Introduction

Do you ever play in the rain and wonder how the water got into the clouds? Have you watched water flowing in a river and wondered where it comes from and where it goes? Have you ever noticed that ocean water is salty and lake water is not and then wondered why this is so?

Water is very important for life on Earth. Without water, life could not exist. Geologists study water and how it travels around the Earth.

7.2 The Hydrosphere

The hydrosphere is the name for the water part of Earth. All the water on Earth makes up the hydrosphere. The hydrosphere includes all the water in lakes, rivers, and the oceans. It also includes rain, ice, snow, and the water in clouds and in the ground.

Water exists in three forms—as a liquid (flowing water), as a solid (ice and snow), and as water vapor (in the clouds). Part of the way water moves around the Earth is by changing from one form to another. Liquid water in oceans, lakes, and rivers evaporates, or changes from its liquid form to water

vapor, which is water's gaseous form. When liquid water freezes to become ice and snow, the water changes to its solid form. When ice and snow melt, water returns to its liquid form.

7.3 The Water Cycle

The way water moves around Earth is called a cycle. Recall that a cycle is a series of events that repeat. The water cycle begins when liquid water flows from rivers into the oceans and then evaporates. Evaporation puts water into the atmosphere where it forms clouds. Then rain puts the water back into the rivers. The cycle begins over again when this river water flows into the oceans.

7.4 Earth Is a Water Planet!

How much water is on the Earth? If you look at a globe or a map of the Earth, you will see that oceans cover most of the planet. In fact, oceans cover almost 3/4 of Earth's surface!

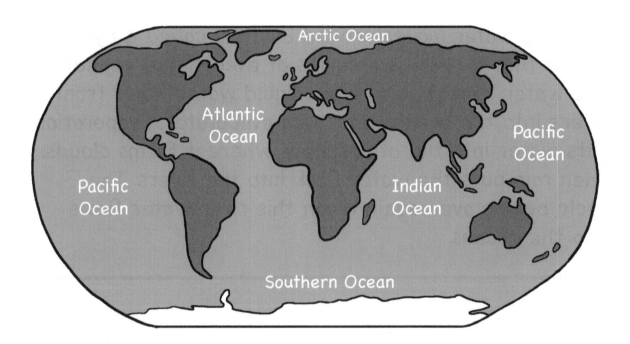

Oceans are very important for life on Earth. In addition to providing evaporated water that can fall as rain or snow, oceans absorb heat from the Sun. Ocean water is constantly moving around the Earth and carries this heat around the globe, keeping Earth from getting too cold.

The oceans gradually release into the atmosphere the heat they got from the Sun. This warms the air above the oceans. Winds blow the warmed air over the land which warms the land.

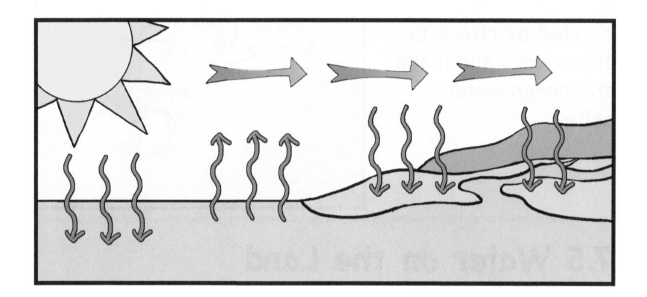

Also, air over the oceans may be cooled by the ocean water. When winds blow this cooler air over the land, the land is cooled and kept from getting too hot. The oceans have a big effect on temperature and weather on the Earth.

Even though the oceans contain so much water, the water is not good for drinking—it is much too salty to drink. The salt in the oceans comes from rocks. Tiny bits of rocks are worn away from big rocks by rain, wind, and moving water. These very tiny rock bits are carried by rivers to the oceans and make the ocean water salty.

7.5 Water on the Land

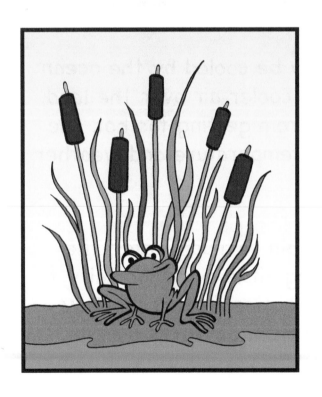

Surface water is the name for the water that is on top of the land. Surface water is found in lakes, rivers, streams, swamps, and marshes. Surface water is very important for life. Animals and plants would not be able to live without water.

7.6 Water in the Ground

Water that is under the Earth's surface is called groundwater. When it rains or snows, some of that water seeps down into the ground. Plants can take up this water through their roots and use it to stay alive and to grow.

There are some places in the ground that hold lots of water. Wells pump this water to the surface where it can be used for drinking and other purposes, such as farming.

7.7 Keeping Our Water Clean

The same water is used over and over on Earth. The same water will sometimes be in the oceans, sometimes on the land or in the ground, and sometimes in the atmosphere. The same water keeps changing between its liquid, solid, and gaseous forms and keeps moving around the Earth.

Without clean water, life could not exist on Earth. Since we use the same water over and over, it's important to keep it clean. But people are not always careful to keep our water clean. They throw trash and chemicals into rivers and the oceans.

Also, smoke from factory smokestacks and exhaust from cars can dirty, or pollute, the air in the atmosphere. This pollution mixes with water vapor in

the clouds and falls to Earth as rain or snow. If there is enough pollution, it can make plants and animals sick.

Geologists and other scientists are studying pollution to find out how we can change the way we do things so that we can keep our water and air clean.

7.8 Summary

- The water part of Earth is called the hydrosphere.

- Water moves around the Earth in a cycle, or series of events that repeat.

- Oceans hold most of the Earth's water.

- Clean water is important for life to exist.

Chapter 8: Plants and Animals

8.1 Introduction

Earth is not made of just rocks, water, and air. Earth has trees, frogs, butterflies, rabbits, deer, and worms. Earth is the only place we know of that has living things. The living things on Earth make up what is known as the biosphere.

The biosphere contains all living things and every place where life can exist on Earth. The biosphere includes plants, animals, and bugs, and also the land,

the water on the land, the oceans, the part of the atmosphere near Earth, and even some underground places.

Different parts of the biosphere work together to help support life. For example, the soil provides water and nutrients that plants need to live. Animals use the plants for food and drink water. Birds fly in the atmosphere to catch bugs for food. Plants and animals get the carbon dioxide and oxygen they need from the atmosphere.

8.2 Cycles

Water is not the only resource on Earth that is used over and over again. There are also different elements, or atoms, that are used repeatedly by living things in the biosphere.

The elements carbon and oxygen are used over and over again in a carbon-oxygen cycle. Think about the oxygen we and other animals breathe in from the atmosphere. We inhale oxygen atoms that we use to power our bodies. We breathe out carbon dioxide.

Plants use carbon atoms from carbon dioxide to make food and then release oxygen atoms back into the air where animals breathe them in again. Oxygen atoms are used over and over, and carbon atoms are used over and over in the carbon-oxygen cycle.

Nitrogen also has a cycle. In the nitrogen cycle, nitrogen from the atmosphere goes into the soil where

bacteria change the nitrogen into a form that plants can use. This process is called "fixing" the nitrogen. Plants absorb the "fixed" nitrogen with their roots and use it to grow. Animals eat the plants and use the nitrogen from the plants to make proteins and run the machinery inside their cells.

Without the carbon-oxygen cycle and the nitrogen cycle plants and animals would not be able to live.

8.3 The Sun

Do you know how animals get energy from the Sun?

By eating plants! When sunlight shines on plants, the plants use the sunlight to make sugars to use for their own food. When animals eat the plants, they get energy from the Sun by using the sugars that the plants made from sunlight.

8.4 Environment

An environment is everything that surrounds a living thing in the area where it lives. Water, weather, soils, plants, and animals are all part of an environment.

Scientists study how all the different parts of an environment affect each other. How much water do the plants in a particular area need to have in order to grow? Which plants will certain animals eat? Which

living things exist in environments that are hot and dry? Which ones live where it is cold or wet?

Learning about different environments helps scientists understand what resources are needed for plants and animals in a specific area to grow and be healthy.

8.5 Summary

● The biosphere is the living part of Earth and contains all the living things on Earth.

● Living things in the biosphere use oxygen, carbon, and nitrogen atoms over and over.

● An environment includes everything that surrounds a living thing in the area where it lives.

Chapter 9: Magnetic Earth

9.1 Introduction

Have you ever noticed how a refrigerator magnet sticks to metal things? Have you ever played with two magnets and observed how two ends of them will stick together and other ends will not? Have you played with a compass and observed how the needle always points in the same direction? All of these events occur because of magnetic forces.

9.2 Magnets Have Poles

A magnet is a particular kind of metal that can create magnetic forces. Magnetic forces allow a magnet to attract certain types of metals to it. Magnetic forces surround a magnet in what is called the magnetic field.

Magnets are said to have poles, or opposite ends. The poles in a magnet occur because the magnetic forces are going in opposite directions. The north pole of a magnet is where the magnetic field points outward, and the south pole is where the magnetic field points inward.

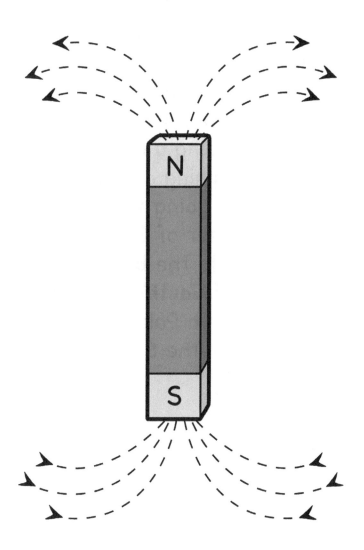

If you have two magnets, you can find out which of their two poles are the same and which are different. The poles that are the same will repel each other (push each other apart). The poles that are different will attract each other and stick together.

You can learn more about magnets and how they work in *Focus On Elementary Physics*.

9.3 Earth Is a Magnet!

North Pole

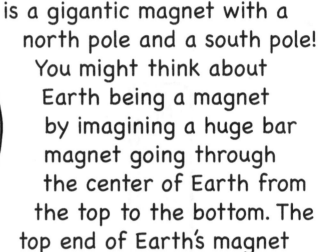

South Pole

It's hard to believe, but Earth is a gigantic magnet with a north pole and a south pole! You might think about Earth being a magnet by imagining a huge bar magnet going through the center of Earth from the top to the bottom. The top end of Earth's magnet is at the North Pole and the bottom end is at the South Pole.

We learned in Chapter 4 that the outer part of Earth's core is made of molten iron and nickel. Scientists think this molten part of Earth's core swirls around, creating a magnetic force. This magnetic force surrounds the Earth in a magnetic field.

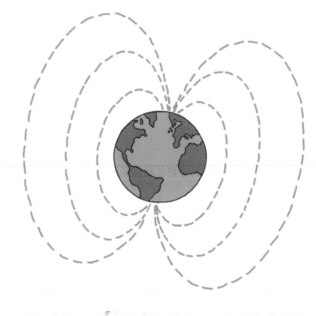

Did you know that you can use Earth's magnetic field to find your way out of the woods? When you use a compass, the magnetic needle in the compass is attracted to the Earth's North Pole, so the needle always points to the north.

9.4 Earth's Magnetic Field in Space

Earth's magnetic field extends into space and is affected by heat and light energy sent out by the Sun. This energy is called solar wind. Earth's magnetosphere is formed when the solar wind hits the magnetic field.

Plants and animals need heat and light energy from the Sun in order to live and grow. But too much of this energy would be harmful to life. The magnetosphere protects life on Earth by letting just enough energy get through. The excess energy is stopped by the magnetosphere. This extra energy then slides around the magnetosphere and continues on into space.

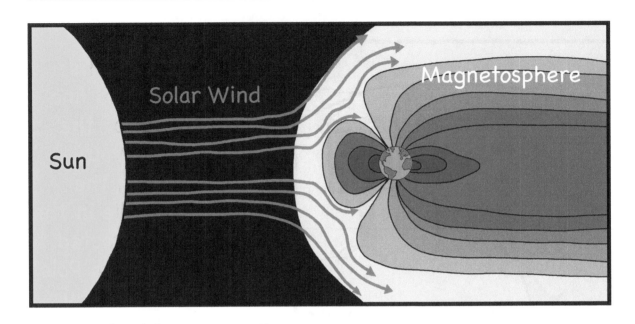

9.5 Summary

● A magnet is a certain type of metal that can attract certain other metals.

● Magnetic force allows a magnet to attract other metals.

● A magnet has opposite poles called the north pole and the south pole.

● Earth is like a giant magnet.

● The magnetosphere contains Earth's magnetic field and protects Earth from getting too much energy from the Sun.

Chapter 10: Working Together

10.1 Introduction

Did you know that all the parts of Earth depend on each other? Without the atmosphere, plants and animals in the biosphere could not get the oxygen and carbon dioxide they need to live, and there would be no rain to bring them water. Without water from the hydrosphere, the cells that make up plants and animals could not produce oxygen and carbon dioxide to go into the atmosphere.

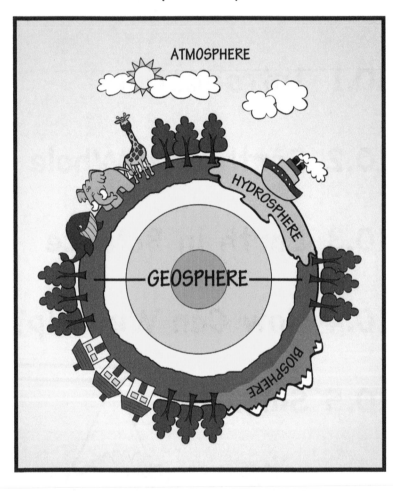

Without the swirling of iron and nickel in Earth's geosphere, there would be no magnetosphere. Without the magnetosphere, plants and animals would get too much energy from the Sun and would die.

10.2 Earth as a Whole

All of the parts of Earth fit together in just the right way, like a big puzzle. In order for Earth to function as a whole, it needs all of the pieces to be in place.

The geosphere, biosphere, hydrosphere, atmosphere, and magnetosphere all work together to make up what we know as Earth. All of the parts of Earth depend on each other. Just like you depend on members of your family and members of your community to grow and live, all the parts of Earth depend on each other to keep Earth working. If you were to take away any one part, Earth as we know it wouldn't exist.

10.3 Earth in Balance

Earth's parts are in balance with each other. There is enough liquid water and water vapor for rivers, oceans, clouds, and rain. There are enough plants to produce oxygen for animals and enough animals to make carbon dioxide  for plants. There is enough of the Sun's energy for plants to grow and a strong enough magnetosphere to block excess energy from the Sun.

However, it is possible to throw Earth off balance. If too much carbon dioxide were in the atmosphere and there weren't enough plants to change it to oxygen, the Earth's climate would change. As a result, the Earth would become too warm or too cold. If too much liquid water were stored as ice, there would be less water in the oceans and they might not be able to support life. If too much ice melted, weather patterns could change, making some parts of Earth too wet and some parts too dry, some parts too hot and some parts too cold.

10.4 How Can We Help?

Keeping Earth in balance is important for life. Many of Earth's cycles can adjust to small changes, but if the changes were to get too big, Earth's cycles could begin to work differently from the way they do now. Scientists don't understand everything about how Earth's cycles work, and they don't know everything about how to keep Earth in balance.

Humans can both help and hurt Earth's balance. For example, humans make some chemicals that can create problems for plants and animals. If too many chemicals are in the environment, plants and animals can die. But if humans clean up the harmful chemicals and replace them with ones that are not harmful to living things, the plants and animals will have a better chance of staying healthy.

Humans also use products, such as plastics, that can create problems when a lot of them get into the oceans. Scientists are trying to discover how to make materials that could be used like plastics but would be changed into harmless substances after being used. This would be a great step toward keeping the oceans clean.

Scientists are working on many new ideas that could help keep our planet healthy and in balance. Maybe you will come up with the next great idea!

10.5 Summary

● All of Earth's parts work together.

● The atmosphere, biosphere, hydrosphere, geosphere, and magnetosphere all depend on each other.

● Earth stays in balance naturally.

● Human activity can change Earth's balance.

CPSIA information can be obtained at www.ICGtesting.com
Printed in the USA
LVOW02*0811190813

348538LV00001B/1/P